Basic Video Production

Everything You Need to Know to
Make Video Like a Professional

By
G L Franklyn

ISBN: 979-8-9876297-5-8

Basic Video Production

Everything you need to know to make Video like a professional!

By G L Franklyn

Table of Contents

Introduction

My name is Gregory Franklyn and that is possibly the least important thing you will read in this book. I'm a guy who accidentally stumbled into a magical world of technological mumbo jumbo that changed my life forever. I owe it all to a guy named Jim Spagg. That's probably the second least important thing you'll read here. But it's a cute story.

I was watching TV on an unusually boring evening in 1991 when, flipping through the channels on cable like we all used to do, I saw an image on the screen that I could NOT believe was on TV. I was looking at a beautiful scene of a meadow and off in the distance was a wide stand of brush or bushes or something. Out of these bushes comes this middle aged short, round, not particularly attractive man whose hairline was receding so fast it was leaving skid marks. He comes out of the brush wearing a pair of combat boots and flailing his arms like a man possessed. Picture this, the guy was wearing combat boots and I mean ONLY combat boots.

I'll spare you the details of what that image looked like because one of the few regrets I will have at the end of my life will be that I was subjected to that image. There's just no way to wash that out of your mind. Talk about Impact! He runs up and puts his face right in the camera, makes this goofy face and says with a maniacal grin, "If you don't like what's on TV, come down here and make your own!"

Over the ensuing years I grew to abhor everything Jim Spagg was about, but he was right about one thing. Video has a HUGE impact. Not just on me, and that image I'll live with for the rest of my life, but it has changed everything for nearly every living soul on the planet. It continues to do so, albeit in a way that makes me nervous these days. However, there is no doubt in my mind that video will continue to change the world and YOU, it would be hoped, will be one of the people that will change it.

Jim was a PEG television producer. PEG stands for Public, Education and Government television. Not to be confused with PBS or Public Television. Public Television is restricted every bit as much as commercial broadcast television, and indebted to the very same forces of money and power. PEG television, and now platforms like YouTube, is YOUR television. It is the last bastion against fascism. Without it, all media is essentially controlled by about 6 boardrooms at multinational holding corporations. I'm pretty sure I don't have to tell you what their interests are.

The airwaves that transport television and radio signals to your home belong to you, not broadcasting networks and multinational corporations. The FCC was established to insure that your best interests would be served by their use. Stop laughing, this isn't funny!

Back when our country was young there were few enough of us that we could all gather together at the church or the fountain at the Town Square to speak our minds and help to make decisions for our communities. Today our sheer numbers and diverse lives make that impossible. But, the need for us to communicate with one another is even greater now.

Television, and now YouTube, is the new town square. It is the essence of how we communicate with each other as a society. It's no secret that the FCC has long since abandoned its mission in favor of establishing a rough estimation of a training monopoly. While watching television, most viewers have no clue about what they are being trained for. But the grocery stores owned by those same corporations know, as does Big Pharma, insurance companies, Auto Manufacturers and on and on. Make no mistake about that! They know what they are training you for.

Back to Jim Spagg. My response to what I saw and what he said was. "I KNOW I can do better than that!" So I did. What a ride it's been. I learned everything I could, volunteered on everybody's shows, produced a mountain of programs and ended up on the staff of the station as the Public Access Coordinator. I taught classes for about 5 years in the subjects you'll read about here. My main function was to empower you to say what you have to say whether I like the message or not.

One of my favorite movies is "The American President" starring Michael Douglas. In the big speech at the end of the movie the president decides to campaign in earnest against his opponent. He says something I think is profound and, pretty much, sums up something that is deeply important to me. He says. "You want free speech? Let's see you acknowledge a man whose words make your blood boil, who's standing center stage and advocating at the top of his lungs, that which you would spend a lifetime opposing at the top of yours…. Now show me that, defend that, celebrate that in your classrooms. Then you can stand up and sing about the land of the free!"

I am a stand for your right to use this powerful tool to change the world and this book is my attempt to give you the basic tools you'll need to get started.

That's the background I come from, but the preaching and evangelism ends right here. This book is about the nuts and bolts of basic video production theory. With this information you should be able to say what you have to say, whatever that might be, in a way that can have an impact and communicate with an audience effectively.

Most of what you'll read here is less about operating the equipment you'll be using and more about what to do with it to communicate effectively. The wheels and cogs of video equipment is an issue for another day. The equipment being used is changing rapidly and a "How To" book about equipment would have a shelf life of roughly 45 minutes before becoming obsolete. The theory of video production, however, does not change and that's what I'm attempting to offer here.

Of course I hope you change the world, but even if you have no such interest, I hope you go out there and contribute what you ARE interested in. From my experience, you have no idea the lives you will change. You will likely never know those people or the impact you'll have on them, but it's there anyway. Make it something that matters to you.

The Image

Shot Composition

I'd like you to watch television a little differently for a few hours. You're looking at it with another set of eyes this time. I want you to notice things that you likely have never noticed before. Does the picture look centered? Is there anything distracting your attention from what's being said? If you're seeing a person, is the picture cutting off part of that person's face while leaving too much room above his head? Where are the people in the picture standing? If there's a graphic telling you who the person is, does the graphic cover part of their face or obstruct anything you want to see? Are they in the center of the picture? Are they off to the side?

These are all elements of what is called "Shot Composition" and if you've ever seen home movies you've probably seen BAD shot composition. Bad Shot Composition will distract the attention of a viewer like no other element of video. The reason being that your viewer's concentration will be split between what the message is, and trying to mentally adjust the camera angle so that what they're seeing looks natural. Bad shot composition detracts from your message.

A few of the stories on your evening news are a good example of what I mean. When your news anchor is about to show you a home video of some event, they will always prep you for what you're about to see. They will say that what you're about to see is a home video to prepare you to expect it not to look right. What they're doing is trying to tell you not to try and compensate in your mind for the shortcomings of shot composition so that you can get the message without splitting your concentration between the awkwardness of the image and what's actually happening.

The Rule of Thirds

Fortunately for you, there's a simple rule to use to get good shot composition. It's called "The Rule of Thirds" and it comes from your childhood. Mentally draw a tic-tac-toe grid on your TV screen while you're watching TV. Equal parts just like you're about to play Tic-Tac-Toe on TV.

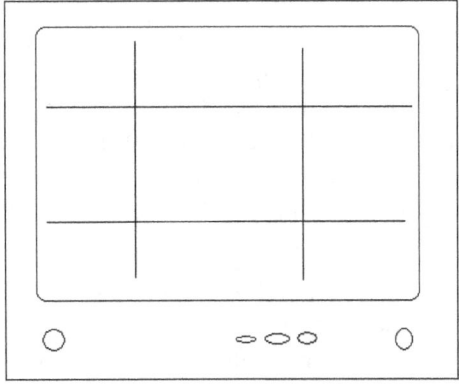

Notice that the subject's eyes will always be somewhere close to the top line and if the subject is facing you directly, they'll be in the center block of the grid. If they're facing left. They'll be somewhere close to the right line and when facing right, somewhere near the left line. That left/right composition is called "Nose Room".

Nose room is needed if a person is not facing forward because if you don't allow nose room the subject will look like they're talking to or looking at the side of the screen rather than talking to someone or looking at something the viewer can not see. Basically what is happening is that you are leaving room on the screen for what your viewer is expected to fill in with their mind. They hear a voice so they naturally fill in a person that they can not see or the subject is looking at something so the viewer needs room to fill in what the subject is looking at.

But wait a minute, Gregory, in that very same newscast, the anchor facing the camera but is positioned on the left side of the screen! (I can see you're going to be a problem in this class, go sit in the back and keep quiet.). True, the anchor is on the left, but there is another element to the picture. There's a little box on the screen with some video, or more often a cute little graphics box, above the anchor's shoulder. Those little boxes are called "Fly Ins, "Cut Ins" or Overlays" and shot composition requires you to compensate for their presence by moving the anchor to the left. The object here is to incorporate the two and center the big picture to include both. You'll notice on your grid that the box is on the right line and the anchor is on the left line (Usually, anyway, sometimes it's reversed.)

The last element on your tic-tac-toe grid is called "Headroom". Headroom is the distance between the top of a subject's head and the top of the screen. Depending on how much of a close-up you're doing, you should watch the headroom carefully. The closer the shot, the less headroom is needed, the longer the shot, the more headroom. You'll notice that the eyes generally remain close to the top line no matter how close or long the shot.

If your shot is not about a person, you'll want to put the most important object to the scene in the center square. A good example would be that groundbreaking shower scene in "Psycho". In one of the shots you can see a little bit of the actresses lower leg, but the shot was centered on the blood going down the drain. The most important thing in the shot wasn't the actress, it was telling you that she was being murdered. That scene is taught in film schools everywhere for good reason. It is a stellar example of shot composition and telling a story effectively. The actress was never injured, but you will likely never forget seeing her murder. Alfred Hitchcock focused your attention on the horror of what a murder would be like without actually murdering anyone. He simply implied with a succession of quick shots that the murder was actually happening by focusing your attention on the little details AROUND the murder and your mind simply filled in the rest of the action.

So the rule of thirds also applies to subjects that are not people. The center square is for single objects you want the viewer to focus on and the rest of the grid is for more than one object. You want to center the whole scene using the grid as a guide.

Backgrounds & Settings

Another element of shot composition is Background or Setting. I'll use a newscast again because it's a good point of reference and then move on to other uses of background settings. While you're watching your news broadcast and a reporter is interviewing a subject notice what else is in the picture. If the subject is a lawyer, for example, you'll see him in front of a bookshelf insinuating he's in a law library. If the story is about a court case, the subject will be in a courtroom or on the steps of a courthouse. If the subject is a fire chief, you'll see them in front of a fire station or fire engine.

Backgrounds help to get the message across by putting subjects in an environment that tells the viewer what the scene is about. Another example might be the "Man on the Street" shot. If the reporter is doing a live story, and they interview a witness to a crime, you'll see that witness in front of the crime scene. There will be at least one of those yellow police caution tapes visible and probably a long shot of the crime scene behind them. If the story is about a political figure, you will see them at a desk or sitting or standing in front of a US Flag, or a congressional building or the White House or some other setting that says, this is about politics.

When choosing a background for ANY scene, place the subject in an environment that enhances the viewer's understanding of what they're seeing. This will help to keep your viewer's mind on the subject. Imagine seeing an interview with a tennis player standing in front of a group of people in a knitting circle. Your viewer is likely to be confused by that. Right there you're splitting the attention of your viewer by making them try to figure out what knitting has to do with tennis while you are trying to focus their attention on the tennis player's story. "But, Gregory, the story is about a tennis star that is teaching handicapped people how to knit!" (Wait, didn't I tell you to sit down and be quiet?)

In other types of video shoots the same rule applies. If you're shooting a love scene, for example, you wouldn't put your subjects on a roller coaster, or at a horse race unless you're trying to convey a sense of sexual urgency or trying to show that passion shows no regard for setting. Like the hundreds of scenes you've seen of lovers making out in the back seat of a car. You'd normally put them in a quiet setting with low lighting in a bedroom or in a living room on a couch or in front of a fire.

There's an exception to every rule. Settings are a good example. If you're shooting an action adventure, just about ANY setting will work, but it will work only because the focus of the scene is the chase, or the explosion or the action and, in that case, where the action takes place isn't relevant to getting the message across. In this case it's about the action, not the characters or where the action is taking place.

One more important thing about settings would have to be "Establishing Shots" or what videographers call "B-Roll". B-Roll is shots of a location, usually, without any people or action used for transitioning the viewer from one aspect of a story to another. When you're watching dramas, or particularly situation comedies, you'll see quite a bit of B-Roll to establish the location of the coming scene. Hence, "Establishing Shots".

A good example, and I'm sure you've seen this done, is when a scene in a show at a workplace ends and the next scene takes you to the subject's home. You'll see an outdoor shot of the residential street or the house in question that tells you that you are now watching something that takes place in another location at a different time. Think 90's comedy series "Rosanne", everybody knows that the Connors live in a house at the corner of 3rd & Delaware because you've seen that street sign a BUNCH of times even if you've only seen a few episodes! That street sign is a classic establishing shot. When you see that sign you KNOW the next scene takes place in the Connor home. The building where "Will & Grace" live is another good example. You've probably seen it a hundred times and you know what it means. Establishing shots give your viewer time to make the transition with you and setting them up to better understand the context of what they're about to see.

Viewer Tolerances

In PEG television there's a well-known acronym called LBM TV. It refers to City Council meetings, State and National Legislature hearings, School Board Meetings; many of which are broadcast, or more accurately "Cablecast" on TV. These programs require a certain type of media viewing. The purpose of these programs is not or entertain or to deliver a particular message. Their purpose is more along the lines of recruiting community involvement and keeping government and education more open to the public. LBM TV stands for "Long Boring Meeting TV". This type of television is critical to the health of a Democratic Republic. Some find the term offensive or demeaning. Please don't be one of them! Everyone who works in television knows what it means and that's why it's used. It's an inside joke meant to denote the downside of producing it. For the people making this kind of video, there's almost nothing to do during shooting except stand there and make sure the camera doesn't get knocked over or burst into flames or something.

This is a perfect example of violating the rule of viewer tolerance. Unless you have a vested interest in what is being discussed, you are NOT going to be watching these programs even though you really SHOULD. They are hard to watch because they flagrantly violate viewer tolerance. C-Span is a good example of a whole channel devoted to LBM TV. The only time most average folks watch C-Span is when there's juicy issues being discussed. You'll remember the Senate Confirmation Hearings for a couple of Supreme Court nominees because so much was riding on the outcome. Justices Brett Kavenaugh and Clarence Thomas come to mind. The hearings were interesting ONLY because we had a vested interest in the outcome.

This illustrates the core element of Viewer Tolerance; Attention Span. The length of time an average viewer will continue to focus their attention before getting tired and beginning to think of other things. In that regard, I'm pretty average. Some will be able to focus much longer, others maybe not so much. When the Comedy Central network first came on cable, their most popular show was called "Short Attention Span Theater". I found it both funny and profound. We are living in a world that is moving progressively faster. Not just because time accelerates as you age, (I'm a little bit younger than dirt. It's moving at lightening speed for me!) but also because technology has accelerated the pace of life in general. The average attention span changes with that pace.

Currently, the attention span of an average adult is about the length of a hit song on the radio. That's why hit songs rarely last longer than about 3 and a half minutes. Attention span and pacing are critically important aspects of viewer tolerance. If you have a message to deliver that you want people to receive. You're going to have to make some sort of change in what you're showing them at least every few minutes. Make a change in camera angle, perhaps a change of subject or maybe introduce a new element of the message. These changes constitute the "Pace" of your program. When you're shooting footage, keep in mind that you will be editing that footage. Make sure you change something every few minutes, or in some cases every few seconds, so you can control the pace of your program when you get to editing.

Keep things moving along at a pace that makes sense for the subject being observed. If you're doing a program about basketball you're going to need to keep it brisk because viewers are accustomed to the speed of the game. You'll need to change your camera angle or introduce new elements every few seconds to keep up with the pace of the game. By the same token, a program about baseball or golf can move a lot slower. Viewers will not be distracted by a lackadaisical pace, in that case, because of the pace of the game itself. However, always keep it moving and avoid long stretches of ANYthing that is not essential to the message.

Tolerances for Comedy are pretty short. That's another reason why the title "Short Attention Span Theater" is so funny. You'll rarely see a scene in a comedy that lasts longer than the average attention span. Dramas and movies have a little more flexibility, but not much. To test me on this, try timing a few sit-com scenes and tell me if I'm wrong. If you're going to compete for people's attention you have to make your message attention friendly.

YouTube now has a huge impact on community media. I love the fact that the playing field for video communication has been leveled to the point that it is truly egalitarian. It is open to everyone regardless of income, social status, gender, race, color, creed and/or (fill in the blank). I suspect that you will, at some point, likely be changing the world using YouTube or something like it. Everything I'm telling you in this book applies to those platforms just like it does for Public Access Television. The only barrier now is the sheer volume of what's available. Just because you post a video on YouTube doesn't mean anyone is going to see it. You'll need something to set yourself apart from the crowd. You'll need the knowledge you're getting right now about how to do that.

Another good example of violating viewer tolerance is this chapter! But, since you have a vested interest in the subject I can go all LBM on you without worrying about it. So anyway,,,,,,,,

Lighting

The Colors of Light

Sometimes you take a picture, any old picture and it doesn't come out looking right. Often it's the color, or it's too dark and you can't make out what it is. Other times it's too bright and the image is what we call "Bloomed", or big spots of white where too much light was pointed at the subject. All of these elements are lighting issues. Different types of light have different colors or "Temperatures" as they are called in video production. It is often difficult for your eye to tell the difference, particularly if your not actively LOOKING for it, but a video camera sees it very clearly.

For example a fluorescent light to a video camera has a slight green color to it. An incandescent (standard light bulb) light has a slight gold or red tint and Sunlight is slightly blue. Halogen is about the only light that a camera thinks is white or natural to reflect actual colors of objects. They each have a temperature of their own. Red would be the lowest Temperature while Purple would be the highest. Ring lights used in YouTube Videos now have a few basic settings for matching color temperatures and can almost eliminate the need for this entire chapter. Almost.

My horror story about learning the importance of lighting temperatures was a scene in a documentary I did years ago (and got a juried award for it, thank you very much) about a close friend who is a very good art photographer. I was interviewing him in his photo studio and he was sitting by a window with sunlight showing down on him. Well, there was also an incandescent light on in the room because he had just taken a break from a shoot. He was saying something really profound and we couldn't take the chance that it would come out the same if I made him move to another spot and say it again, so we had to use it. The horror of it was that the light bulb made one side of him really red to the camera and the sun made the other side of him blue. He looked like a circus clown, but what he had to say was just too important to lose.

Needless to say, I went back to the station and took an advanced class in lighting, which I eventually ended up teaching. I'm giving you this information as simplified as I can. The subject of light gets really cosmic and spiritual when you get really deep about it. But the short version is to think about lighting from the viewpoint of a camera. When you're working in Black & White like my friend did, it doesn't matter much. I think it was in that very interview that my friend confessed that he likes Black& White photography *because* it "Forgives a multitude of sins." It usually doesn't matter to the naked eye, but to a color video camera light temperatures are critical.

Remember the temperatures I gave you above and keep some compensating color gels with you whenever possible. For example, in the situation above, if I had used a light blue gel over the light bulb, that would have equalized the color temperature of the lights and he would have looked fine. Any store that sells lighting gels can tell you the exact shade for balancing sunlight with other sources and it's relatively cheap to buy. Most consumer video cameras automatically "White Balance" for you every time you turn them on. The function of White Balance is to tell the camera what white is so it can determine what all of the other colors should look like. White was the standard cameras used to do that. Back in the day, videographers would use a sheet of plain white paper or a reflector board to show the camera what white is. Now-a-days cameras and cell phones do that on their own. The only time you might run into trouble with color temperature is when mixing different temperatures.

In the case of fluorescent light that's a little more difficult. There is no real balance for green. Sometimes a light rouge gel will make it better. But, your best bet is to try to overpower fluorescent lighting with halogen or some other color. The green in fluorescent lights is not as pronounced as gold/red is from a light bulb. If your camera can be manually white balanced, you should do so under fluorescent lighting only and you'll probably be fine. Your main light temperature challenges will probably be sunlight and any other color and a light blue gel on the non-sun light will usually correct it. Use light sparingly. Most modern cameras don't require much help with light. Intentional Lighting like what I'm describing here is mostly intended for getting a particular effect to enhance your message.

Three Point Lighting

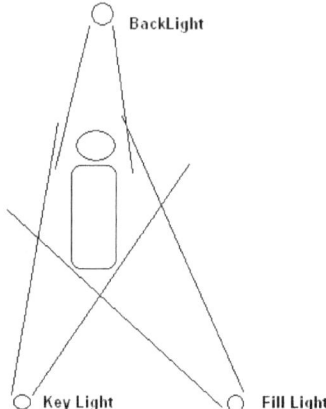

Placing the lights, if you use them, is another flexible tool to use like a background or setting. The most common is "Three Point Lighting". Three Point lighting refers to the three directions from which you cast light on a subject. One is called the "Key Light" and is the main light on the subject. Usually you want that light to come from just about anywhere above eye level and a little to one side or the other, rather than head on. You are approximating the angle of the sun, which has varying heights, but usually it's above you and usually people are not looking directly into it. Slightly above eye level is best so you don't have a deep shadow under your subject's nose or eyebrows.

The second of the three lights is called the "Backlight", and as I'm sure you've figured out already, it comes at the subject from behind and above. This light should have a sharp downward angle. Its main function is to outline your subject, to pull them forward from the background and to soften any shadows on the background caused by the Key Light. This one should be about half the intensity of the Key Light and should be just enough light so the viewer can clearly see where the subject ends and the background begins.

The third light is called the "Fill Light" this one should be about half the intensity of the Key as well. The function of this light is to fill in the things that the key light is missing because of its angle. This light can come from almost anywhere in front of the subject opposite the key light. Most often it's lower than eye level and at a triangle location from the Key and Back Lights.

Many good videographers use a "Light Board". A light board is any flat object that will reflect the backlight back on to the subject. Sometimes they're a big cloth disk that folds up for easy transport and has a reflective silver surface on one side and a clean white surface on the other. The two sides determine the intensity of the light being reflected back on to the subject. Mostly used in sunlight environments. This reflector is a good example of "Cheating" with light. There really isn't a fill light, you're just making it LOOK like there is. Cheating is handy for tight budgets. You can even use a white or light colored wall if you like, just be sure to watch the color of the reflection. Remember my "Clown" story.

Lastly, the fourth light in three point lighting (see what I did there?) is a light behind the subject pointing at the background. The intent of this light is to eliminate any shadows on the background that may be caused by the Key and/or fill light. It can be added any of the lighting configurations I'm describing here, as needed. Wait a minute, Gregory, if it's called "Three Point Lighting" why are there FOUR lights? (Hey, I thought I told you to sit down and be quiet!) The 4th light is hardly ever needed, but have that in your arsenal just in case.

Two Point Lighting

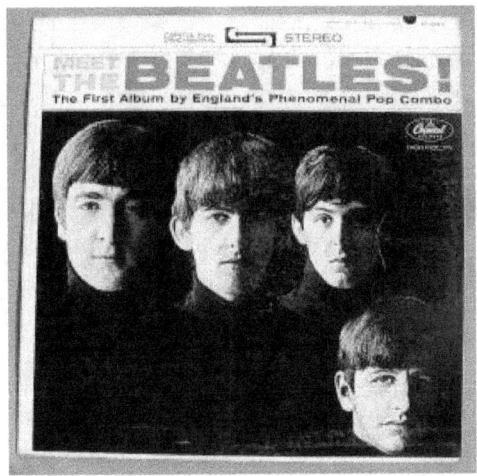

Two Point lighting is three point lighting without a fill light. Your objective with this type of lighting is to give your subject a dramatic feel. A good example of good two point lighting that you've probably seen is the cover of the very first Beatles Album from back in the early 60's Almost everyone alive has seen that image. 4 guys with bright light cast from the left and no light from the right, giving them a quiet and subdued look like they were poets or actors. The uses for two point lighting are few and usually involve art of some kind. Two point lighting is all about drama.

One Point Lighting

You are also familiar with One Point Lighting and it has a narrow field of uses. One Point Lighting is the use of the Key Light with no other lighting. A perfect example of One Point Lighting is holding a flashlight to your face from just under your chin. Just like when you were a kid, this effect is designed for spookiness and it works perfectly for those scenes. Depending on the direction of one point lighting you can insinuate a lot of things. Headlights from a car, for instance, a candle or a campfire. Limited uses, but it creates atmosphere really well. Today's Ring Lights that are used in YouTube videos are LED and most folks use that by itself. The light is well diffused and close enough to the subject that other lighting is rarely needed. But you'll be prepared in case you see any distracting shadows through your camera.

Flood Lighting

Flooding light is another type you'll come across a need for. This type is usually an intense light with no particular direction. The purpose of this type of light is not so much to illuminate a subject as much as it is to raise the overall light level of an entire area so a camera can see it better. The most common types of flood lighting is,,, Say it with me now,,, That's right! A Flood Light. Most videographers use what are called "Scoops" and "Totas". They are much better than flood lights because they are truly directionless where as flood lights are more like automobile headlights. They're not really focused beams, but they don't distribute light evenly over a wide area, either.

Use only enough light to get the job done. Unless you're doing a Las Vegas Show number, you'll end up with a cheesy looking finished product. But, who knows, sometimes cheesy is good. The fact is most modern cameras don't require very much help with light. Again, intentional lighting is more for creating specific effects.

Editing

Digital editing is quickly becoming the norm at a time when there is a huge segment of the industry still using expensive analogue equipment. In this section I'm not going to spend much time talking about the operation of equipment used for editing video like I would If I were teaching you how to use it. Here, I'm going to talk about editing theory. What basics you'll need to know to get your point across to an audience

The most complex and mentally challenging aspect of video production is editing. It includes a whole array of elements that need to come together simultaneously in a way that gets your message across without disturbing or distracting your viewer from what you want them to see and hear.

Unless disturbing your viewer is an important part of what you are trying to convey, avoid doing so at all cost. If you're doing a program about cruelty to animals, for example, you're likely to *want* to disturb your viewer to demonstrate how cruelty to animals effects animals. If you're doing a scary movie, you'll want to disturb your viewer quite a bit. Other than situations like that, you're working against your own interests if you disturb or distract your viewers with poor editing choices.

One of the reasons that most people hate to sit through someone else's home movies is a good example of bad editing. The want-to-be movie maker is going to have to explain everything you're seeing because the story would otherwise be impossible to follow. Actually, there usually isn't a story there anyway. It's just a bunch of raw footage of things someone else thinks is significant. That type of video has its value to the people who made the video and their friends and family. It is critically important to them, usually because it stirs memories that are important. To everyone else it's useless drivel. Editing is taking that same footage and arranging it in a way that a viewer can understand what they're seeing and hearing without a running commentary explaining it.

Logging Footage

The first thing I want you to remember about editing is "Logging Footage" or whatever format you're recording in. Logging footage is writing down what is on the recording, and where it is located. This will save you a virtual mountain of time when you're editing. While shooting, you're likely to record a lot of footage at a particular location even though, in the finished product, that location will be appearing at different times in the program. Having the recording logged will help you find that footage quickly. Nothing is more frustrating to an editor than knowing exactly what comes next in a program and then having to slog through hours of footage to find it.

Now that digital editing is the norm, particularly for the modern editor, it's not that critical, but it will save a lot of time anyway. When I log, I usually just list the order of the scenes on a medium. Others list the time on the medium, how many seconds long it is, how many takes of a scene are there and even which take has been selected for the final edit. You will probably not need to get that detailed when you first start editing. However, logging footage will save you a lot of time and frustration. I've included a sample Logging Sheet at the end of the book. You are welcome to copy it and use it to your heart's content.

Transitions, Wipes & Fades

When assembling raw footage into a finished program try to think like a viewer. Most viewers think linearly. That meaning, one thing follows another in a logical succession. There will be lots of times when you'll need to "Transition" your viewer from one subject to another. We've already talked about "Establishing Shots" earlier. This is a good example of transitioning your viewer from one aspect of the program to another.

Transitions can say things other than a change of location. You've seen shows where a character falls into a dream, or a flashback of a time gone by. Usually, this transition is signaled by blurring the image on the screen or making it wave or dissolve slowly into the scene that is happening in the dream. This type of transition is called a "Dissolve" and, in the case of a dream sequence or flashback, is always signaled by a slow transition and normally applying reverb to the audio track. It audio-visually mimics what a person normally experiences when falling asleep.

The most basic transition is a "Cut". Changing from one image to another instantaneously. This transition is usually used to change the camera angle during a scene or to show different aspects of a single scene without showing the whole scene in a big picture. The "Psycho" scene we talked about earlier is a good example of cut edits. You see the whole picture in little pieces that your mind assembles into a big picture. Cut edits show immediacy and keep you in the same place and time, where as other types of transitions, like dissolves, wipes and fades always signal the passing of something like time, a scene or a location.

Another use of cuts would be "Montage" editing. A montage is a succession of short images in rapid succession to show a variety of aspects of a single subject. Commercials are a good example of a montage. You'll see quick flashes of a hamburger, a coke, a family smiling or laughing, some french fries, a restaurant setting, all within a time space of a few seconds. That "Psycho" scene is also a good example. The viewer sees a bunch of images very quickly and their mind connects them together into a whole picture. That is called a "Montage". If it matters, Montage is a French word that, loosely translated, means "Montage".

Another common transition is a "Dissolve" or "Fade". At the end of almost any program you'll see a "Fade To Black". This dissolve indicates to the viewer that something is ending without having to say it or put up a graphic stating so. When you're watching a movie, you KNOW when it ends even when it doesn't say "The End" on the screen like in old movies. Film makers know that you can tell when it's over partly because the story is resolving and partly because they're telling you that something is ending by fading to black.

Another kind of transition is a "Wipe". These are a little more rare in video. They are a sharp line of some shape or design that wipes one image away and simultaneously replaces it with another. Like a squeegee wipes water away leaving a clean surface. It is generally used like a dissolve or dream sequence, to alert the viewer that something is changing and to expect a new element to be introduced.

Lastly, you have the "Split Screen" image. This transition indicates two or more things happening at the same time in different places. It is two or more images on the screen at the same time with different subjects in each image. Often it is used to show a telephone conversation between two or more people separated by distance. It is used to establish the difference in location. Once the separation is established you can use cuts to show the conversation and your audience will be able to follow it even though both images are no longer on the screen at the same time.

Using the wrong kind of transition will disturb or distract your viewer. For example, if you use a cut edit to transition from one location to another, you'll confuse your viewer because you haven't let them know what's happening by giving them a warning. In horror movies you'll often see a character in a dark setting moving slowly or quietly and then suddenly, BAM, there's a monster right in your face without warning. The object in that instance is to shock or scare the viewer. That situation is perfect for a cut edit. Taking your viewer one location to another with a cut edit is distracting and disturbs the viewer. It doesn't give them time to make an adjustment to what they're seeing and hearing. That will be when you need an establishing shot to help your viewer make the transition with you.

Graphics & Text

The next element of editing is Graphics or Text. Your evening news is, again, a good example. News programs are notoriously heavy on the graphics. Wipes and cuts and flying fonts and colors and swooshing sounds and montages all coming at you at an exciting pace. It's mostly eye candy to insinuate that you are witness to excitement, you are where it's all happening, You're up to speed, in the know! Graphics can help to establish a feel or a pace for your program. In movies, you almost *have* to tell your story without the aid of graphics or text. You rarely see graphics in movies or television series'. But, in almost every other type of program, you'll need to use at least some text to help out from time to time.

If you're doing a documentary you'll want to identify any number of people, places or things. You'll want to identify someone who is talking about something when the viewer might not know who the person is or why they might be speaking about a subject. In this case you'd identify the person with a "Font" of their name and who they are or what they do. Earlier, we talked about the "Rule of Thirds". That concept is going to come in handy here too. Fonts for identification should always appear below the bottom line of your grid and should not cover any part of the subject's face.

CNN and C-Span are good examples of good font placement. There is almost always a font on the screen and it is always in the bottom third. This is comfortable for most viewers. They will read it once, or maybe twice if the person talking is particularly boring and then get back to watching the action and listening to what's being said. CNN and C-Span keep those graphics up because they know that people are tuning in and out by the second and viewers will probably need that information constantly. For non-news programs you only need to establish who is talking and why, when the person is seen in the program for the first time. After that you can assume that the viewer already knows, because you already told them.

The type of graphics you use should fit your subject. CNN and C-Span use a strip in red white and blue along the bottom of the screen. Red white and blue, or even red and blue, say this is about America, Politics or something Civic in nature. Sports teams will often use the team colors in fonts in their programs. If you're doing a Christmas program, for example, you'd probably pick Red and Green for your fonts because they are the colors most associated with that holiday.

You'd probably use black and orange for Halloween and pastel pinks and blues for Easter. You get the idea. Using those colors enhances your viewer's understanding of what you're trying to tell them.

Lastly, you may want to use graphics or text as an artistic expression in your program. For example, using huge screen filling words or letters that are semi transparent can have a really cool effect if you're making a music video. A word or two out of the lyrics, flashing or dancing across the screen can help to cement the concept of the song in the mind of the viewer. You might also want to make a program consisting entirely of graphics with only sound as a background. Like a Powerpoint presentation. For artistic purposes, there are no rules for the use of graphics or text. Making them discernible will probably help if you're trying to convey something in actual words.

I tend to use smaller graphics without a background color and put a thin line or 2 between the name of the subject and who they are and just use the same 2 colors throughout the entire program, but that's just my particular taste. Unless you're trying to excite your viewer with eye candy, keep the graphics understated so your viewer has the option of looking or not looking. Don't go for distraction unless that is your intent.

Audio & Mixing.

Sound is also critically important to delivering a message. If you've ever watched TV with the sound off you have some idea of how important sound is to any type of video presentation. Most audio for your epic will be pretty straight-forward. The voices of people talking should match the lip movement of the people in the image. The sound of a door closing or a train going by should match what the viewer is seeing. Most of that is already handled on your tape unless you had some problems with shooting and need to do some looping. However, that's only a small portion of the wonderful tool sound can be to help get your point across.

First on the list of powerful effects that can be contributed by sound, is not forgetting to include ambient, or background, sound. For example, if you're shooting footage in a city park you will want to record some sounds that a viewer would naturally hear in a park. Kids playing, a dog barking, and a breeze through some trees or bushes, are all sounds one would hear while in a city park. You'll want to keep that audio kind of low during the scene but you want it to be there so the viewer's experience is more like actually being there in the park with your characters.

Music

Music can have a *huge* impact on a viewing experience. For example, you've seen scary movies and you know how much a little scary music can elevate your anticipation that something is about to happen. Who can forget the theme music for "Jaws", those two notes alternating back and forth have tremendous power to enhance the fear and trepidation of those gory scenes. Likewise that one violin note screeching over and over in that legendary "Psycho" murder scene added so much to the suspense and horror of the murder.

Once you've seen it, how can you forget that sweeping symphonic theme used in the final scene of "Gone With the Wind". It brings up romantic feelings of loss. It is the same theme used at the beginning of the film, just played a little differently to focus on what you are seeing at the time. The theme is kind of light and breezy at the beginning of the movie, denoting that all is well, the sun is shining and all is right with the world. At the end, that same music is played a little slower and with a lot more orchestration with dramatic swells denoting gloom and doom. It's the same music saying two completely different things. Both, critical to what the movie is saying.

Music alone is a powerful thing that is even more powerful when combined with images. Even in a documentary music can help to cement a point in the mind of a viewer. Most professional TV and movie programs have at least one music supervisor. It's that person's job to find, commission or create, the right music to enhance every scene you see. Including those scenes where there is no music at all. The important thing to remember about the use of music is to select it to enhance the message and never let it overpower the message, the dialogue or narration.

More on Audio Levels

Just like in the recording of the image, sound levels of different elements of the scene should enhance the message without overpowering it.

Also remember to listen carefully to the sound your editing into the program. If your audio is too hot it will sound muddy and cheap. If it's too low, it will be distracting and no one will be able to hear it. One other thing I had problems with early on in my productions is sound levels across different scenes. It's a real skill to keep audio levels consistent throughout an entire program. You'll likely not edit your entire program on the same day. It will usually take several sessions. How do you keep the overall sound level consistent so that some scenes are not significantly louder or softer than others?

The most important thing that has caused problems for me about sound levels isn't the volume. It's called "Trim" or "Input Level". Different sources of audio often have different types of equipment that reproduce them. The level of signal a microphone puts out will not be the same as that for a CD player or a camera and your trim control compensates for that difference. The same volume level at two trim settings will be dramatically different. To achieve consistent audio through an entire program, keep an eye on both controls and you'll be fine. IMPORTANT NOTE: When you're finished editing for a session, WRITE DOWN where your volume, tone AND trim settings were when you last edited and re-set them before you start again! The reason I say that is because I was working in a cable access studio and editing room. As soon as I'm done, another producer would be using the same equipment and would change the settings for their own use. Therefore, the next time I would go into an editing suite, it would be all set up for some other use. So, always write down your settings! (You're most welcome!)

The Finished Product

Where to Show Your Work

Once all your sweat and tears have been shed. All the challenges have been met with consummate skill because you learned everything in this book backwards and forwards, the last step will be what to do with your epic now that it's done.

There are a lot of venues for you to display your work. The "YouTube.Com" website is screaming hot for video productions as I write this book. You have *millions* of potential viewers and it's *FREE*! There is precious little as far as gate keeping telling you you can't post your video up regardless of its content.

There are competitions for the type of program you've made, there's a local PEG television station near you that would love to cablecast such a wonderful and informative programs as yours. Go to your computer and throw together a nice cover for your program and donate a copy to your library. Screen it for a local art theater for inclusion in a film festival. Make sure that local organizations who may have an interest in the subject can have a chance to screen it for their staff or clients. Get a copy to local publications that review videos, they may write about it for you. Most publications are starving for inexpensive copy to fill the pages that they sell advertising for. Write a review yourself and submit it with your review copy. Who knows, they may do a little editing and print what *you* wrote.

Audience Reaction

I've done all of those things, but ordinarily I'll screen it myself for my friends and their friends first. I'll throw a party and invite as many people as I can and serve some food and drink and screen it several times over the course of an evening for them so I can gauge their reaction to it. I'll watch them watching the program and check for reactions to things that I wanted them to get. I can tell from their reactions whether I'm communicating well or not.... Or so it would seem.

I have one last personal story and then I'm done. I worked for many years on a weekly PEG TV series called "Night Scene". It was this crazy sloppy dance show at a local nightclub that featured new music, lots of dancing, a movie review or some "Man on the Street" interviews and a little stage show each week, but mostly people dancing to disco music in this poorly lit nightclub with lights flashing everywhere.

It was torture for me to get a good polished product out of the environment because you could never see anything clearly, it was always so dark and lights were flashing everywhere. I tried and tried to get the producer of the show, who was also the club owner. To let me raise the light level so TV viewers could actually see what was going on.

He refused! He said the main issue was keeping the club open and making the environment conducive for the people who come to the club to dance. After several years of pleading and begging, I finally quit because I just couldn't stand not getting a good professional looking product out of the environment. I thought my reputation as a professional video guy would be ruined by that mess of a program.

A few weeks later I happened to be at work at the PEG station that cablecasts the series and the show came on the TV in the lobby. I was all ready to be embarrassed by it in front of my peers. There were about 5 or 6 other community producers in the lobby at the time waiting to get into a studio or editing suite. One by one they became distracted by what was on TV and every last one came over and found a seat and watched the entire show. They were reacting to what they were seeing and hearing just like we had intended while we were shooting and editing it. They looked like they were enjoying it and told me so when the show ended.

I'll never forget that moment. All the time I was struggling with the technical aspect of the show I had no idea that the technical quality had nothing to do with why people were watching it. For the first time ever, I watched my own show through the eyes of someone who had no vested interest in it what-so-ever. You know what? It's a great show to watch. It's good television, there's always something colorful, something moving, some music happening, then some talking and some information and then back to more music and color and flashes of light everywhere. It was exciting with all it's flaws and awkwardness and lack of polish. It was fun!

I told you this story to illustrate something I learned from that experience and mentioned in the introduction to this book. You'll probably never know the people being affected by your program or what it means to them. But, please make that program anyway. Somebody, somewhere really needs to see it. If you don't make that program, the world won't be changed by it! I wish you the very best in making your program and I hope I've been of some help.

"Fade to Black"

BONUS SECTION!

Directing a Studio Production
By Gregory Franklyn

WHAT A DIRECTOR DOES: The director of a production is like the Captian of a ship. He or she is responsible for giving the commands that actually guide the ship while on it's journey. The producer is the person who creates the program and makes all arrangements for people and other items needed for the production before the production begins. Sort of like the ship's owner, who decides where the ship will go and what it will carry. The director comes into play when a production is actually scheduled. The director should be in close contact with the producer during all phases of planning that involve the execution of a production. Once the production begins, the Director is in full command of the production. Like in the nautical analogy, once the ship leaves port, the owner's role becomes limited and the Captain is in command. Defining these roles, early on, and deciding how they will work together will be to everyone's best advantage.

PRE-PRODUCTION PLANNING: It can not be stressed enough how powerful a tool pre-production planning is to set your program up for success. Setting up a pre-production planning meeting as close to the actual production as feasible is the best way to ensure that your production will look and sound the best it can. Include the producer and all crew people who will work on the production. If you are unfamiliar with a crew member, take a moment to introduce yourself and get to know the person at least a little bit. Good communication with your crew will be essential to the success of your efforts. Once the plans have been made and agreed to, STICK TO THE PLAN!!! Changing the game plan moments before the production begins is a recipe for disaster!

A pre-production planning meeting should focus on the execution of the production, rather than it's content. Content should already be decided by the time a pre-production meeting is held. Elements that are important for a pre-production meeting should be:

- Who will perform which specific duties, i.e Technical Director, Shader/VTR operator, Audio, Character generator, Cameras, Floor Director etc.
- How the program will begin and what duties will be entailed.
- The order in which events will happen in the program.
- Which camera angles each camera operator should be looking for during the program.
- Any roll-ins or other tools that will be used during the production and when they will be used.
- What cues will be needed by talent and when they will be needed.
- Commands and Directing Style: The director should use consistent commands during a

program and make sure all crew members are familiar with what the commands mean.

- What to do if anything unexpected happens during the production.
- How the production will end and what duties will be entailed.

Having everyone on the production clear about these items should set you up for success. If other things occur to you along the way, by all means, include them in the planning process.

ONCE THE PRODUCTION BEGINS: Crew should be assembled and in their positions at least 15 minutes before a program begins. The director should run the crew through any unusual operations that will occur during the program, to insure that everyone is used to how the operation will happen. For example, the director should guide the crew through a few practice runs of the opening of the show. This will help to prepare the crew for the opening and familiarize them with what part each will play. This will also "warm up" the camera people so they are ready the minute the recording or cablecast begins. It's a good Idea to run through the transitions to and from any roll ins or other events as well. This will also give the director the opportunity to describe how shots should be composed during the program and what types of shots he or she will expect from the camera operators.

As the Director, you should be familiar with what will happen and in what order.

ZEN WITH THE SUBJECT: Good directors watch and listen carefully to what is happening on the set and block camera shots accordingly. If, for example, someone is giving a heartfelt personal testimony a good director will be able to sense that and prepare a close up of the subject to bring the audience in for a closer look at the expression on the subject's face. If a subject is particularly animated in speaking style, a good director will concentrate on shots that include whatever gestures the subject is using. If more than one subject is involved in a heated exchange, a good director will focus on shots that include both parties at once. It is very distracting, as a viewer, to see a picture of someone, on screen, who is not speaking, while an important point is being made by another subject who can not be seen. Likewise, if a subject is referring to an object on the set, a good director will set up a shot of the object and get it to the audience quickly. Your job as a director is to anticipate what will happen next and be ready to show it to the audience in its most appropriate form.

DIRECTING COMMANDS: Every director is different, however, the differences will usually be subtle. Sort of like the difference between the King's English spoken overseas and American English spoken here. For example, most directors, when setting up a transition from one camera to the next will give the command "Ready to dissolve to #,,, dissolve to #". I use the command "Ready to FADE to #" because it's easier for me to say. It means the same thing, but it's different than the norm. This is an example of something I would explain during the pre-production meeting or the warm up phase. As the director of a production, you may disorient your crew members by switching back and forth between the two commands even though they mean the same thing. Being consistent helps your crew clearly understand what you mean at all times.

"Ready commands" are very important. A ready command is a command that lets your crew know what you will be doing next. Getting back to the nautical theme, your Technical Director would be like your first mate. When you give a navigational command to turn the ship a certain direction, your first mate is the one who actually executes the turn. The technical director, in like manner, actually puts the image you ask for on the screen. Since television happens with considerably more speed than navigating a ship, Ready commands are not only courteous, but essential. Always give ready commands before commands. Example: "Ready to cut to Camera 2" (Ready command) "Take 2" (Command). When these commands are given in this order, a good Technical Director will put Camera 2 on the Preview monitor for the director, at the ready command, and then to the program at the command. This also alerts camera 2 that his or her image is about to become part of the program and lets them know when they are actually hot. Even further, your floor director will use the ready command to prepare to cue talent to the new camera angle. Your Audio Specialist will also use this command to ready any additional or different audio sources if needed. Here are a few common commands used by directors:

- **TAKE** (also cut) this command is used to transition from one image to the next instantaneously.

- **DISSOLVE** (also fade) this command is used to transition from one image to the next gradually. Both images will be on the screen at one time for a short while. Often this command is accompanied by a speed specification. i.e., "quick" or "slow"
- **WIPE** this command is used to transition from one image to the next with a hard line or shape. Both images will be on screen for a short time. Often this command is also accompanied by a speed specification. i.e., "quick" or "slow"
- **HALF DISSOLVE** (also half fade) this command is used to superimpose one image over another so both images remain on the screen for a longer period of time.
- **SPLIT SCREEN** this command is used to divide the screen between two images, both of which will remain on the screen. Be sure to mention which 2 images will be split.
- **ZOOM IN** (also push) this command is used to instruct a camera to use the zoom to bring the subject closer.
- **ZOOM OUT** (also pull) this command is used to instruct the camera to use the zoom to move the subject further away or open the shot to illustrate the setting in which the action is occurring.
- **PAN** this command is followed by "Right" or "Left" and is used to instruct the camera to move the camera lens in the specified direction.
- **CHEAT** this command is followed by "left", "right", "up" or "down" and is used to instruct the camera to move the camera lens, subtly or

very slowly, in the specified direction so as not to be noticeable to the viewer. It is used as a corrective measure for a shot that is slightly out of composition.

- **HEAD ROOM** this command is preceded by "more" or "less" and refers to the distance between the top of a subjects head and the top of the screen.
- **NOSE ROOM** this command is preceded by "more" or "less" and refers to the distance between the subject's nose and the edge of the screen in the direction the subject is facing.
- **FOCUS** this command is used to instruct the camera to adjust the focus on a subject.
- **HARD FOCUS** this command is used to instruct a camera to zoom all the way in and focus on the subject, then return to the previous shot. Used only when the camera is NOT hot.
- **DOLLY** (also truck) this command is followed by "in", "back", "Right" or "Left" and is used to instruct the camera operator to move the whole camera's physical location in the specified direction.
- **TILT** this command is followed by "up" or "down" and is used to instruct the camera to tilt its lens in the specified direction.
- **BRING UP** (also bring in also fade up also fade in) this command is used to introduce audio or character generator information to the program.
- **LOSE** (also cut also fade out) this command is used to eliminate audio or character generation information from the program.

- **READY FONT** (also ready graphic) this command is used to instruct the character generator to find the page containing graphics for a specific subject, item or image and ready it for use.
- **UNDERCUT** this command is used when graphics or audio are on the screen and the director wants them to remain while a transition from one image to the next is taking place.
- **READY ROLL IN** this command is used to instruct the VTR operator to prepare to start a tape that will be part of the program.
- **FADE TO BLACK** this command is usually used at the end of the program or the end of a segment of a program. It is a universal command that instructs the Technical Director, Audio and Character Generator to simultaneously fade down their respective portions of the program and cues the entire crew that no one is hot.

Directing a production entails a great deal of concentration. You MUST be in control at all times. It is extremely rude to interrupt a director during the course of a production unless the matter directly relates to what he or she is doing at that moment. A director should never tolerate arguments from any crew member during a production. In the unlikely event that this should occur, you are well within your job description to ask the offending crew member to leave the studio area.

The respect with which you treat your crew and producer will be a key factor in your success as a director. Like in the nautical theme, the respect you inspire in your crew is essential to your authority to guide them. Never provoke a mutiny! AND, by all means, HAVE FUN!

www.ingramcontent.com/pod-product-compliance
Lightning Source LLC
Chambersburg PA
CBHW070337290526
45791CB00003B/1372